# A Wolf and His Man

Elizabeth Parr  Leyton Jay Cougar

featuring photos by Jan Ravenwolf

Paw Print Press

# CONTACT INFO 

website~ www.wildspiritwolfsanctuary.org
phone ~ 505-775-3304 Wild Spirit Wolf Sanctuary
e-mail ~ info@wildspiritwolfsanctuary.org



Requests for permission to make copies or to place orders for this publication should be mailed to:

Wild Spirit Wolf Sanctuary
Attn: Georgia Cougar
HC 61 Box 28
Ramah, NM 87321

or emailed to awolfandhisman@gmail.com

Printed in the United States of America.
Printing by Printer's Press, Albuquerque, New Mexico.

Back Cover Photos, Photos: Jan Ravenwolf
Cover and Book Design, Cover Photo, Photos: Georgia Cougar
Additional Photos: Allison Bailey, Angel Bennett, Phil Sonier, Angela Albrecht, Leyton Cougar, Helen Garner, Liz Parr, Solveig Lange, Christine Meier, and Jill Fineberg

Library of Congress Control Number: 2010920808

ISBN 0-9843055-0-6

Parr, Elizabeth
I'm Glad You're Not Dead Journey Publishing 1996, 2000
Coping with an Organ Transplant Penguin Putnam 2001
Vivir un trasplante de organo Paidos, Spain 2002

Elizabeth Parr

Leyton Jay Cougar

Jan Ravenwolf

Georgia Cougar

On The Cover
Leyton Jay Cougar with Raven the timberwolf
at Wild Spirit Wolf Sanctuary
Photo: Georgia Cougar
July 3rd 2008

# DEDICATION

This book is dedicated to all the people
whose lives were touched by my friend, Raven.
~Leyton Jay Cougar

*photo: Jan Ravenwolf*

Photo Courtesy of Jill Fineberg. Author of People I Sleep With. Ten Speed Press

# CONTENTS

photo: Jan Ravenwolf

# PREFACE

The Western states of North America worry out loud about its wolf population. What to do with them? Reintroduce them into the shrinking wilderness? Exterminate them because they kill valuable livestock? Kill them because they steal prey from hunters, especially the mighty elk?

We, meaning humanity, find ourselves, as with so much in 21st century nature, managing wolves. Those lovely, swift creatures that the Natives thought of as equals with them in hunting, the same whose spirits were sacred. "Wolf medicine is strong medicine," a teacher once said to a young Leyton Cougar. That occasion set him on a path to New Mexico, to managing a far flung wolf rescue ranch, and to a meeting with his brother Raven, a timber wolf.

This is a kind of love story between the tall, blue-eyed man and the black and silver wolf that would become his partner, that chose him in order to be able to communicate with humankind.

You, the reader, should follow this story out of doors, leaning perhaps against a tree, straining to hear a distant howl, and then a bright laugh, as a wolf and a man turn a corner and let you in to their world.

We are grateful to the supporters of the Wild Spirit Wolf Sanctuary and to those who have contributed to the production of this book: Alice Boynton, an editor, who kindly made sense of some chronology. Georgia Cougar and Helen Garner for their patience in proof reading, over and over again. Georgia's newsletters from Wild Spirit are its blood stream, without

Raven with Lisa

which it could not live and grow. We are exceedingly grateful to Jan Ravenwolf whose photographs make this little volume worth the read and the cost. Without Jan's contribution, we would have half a book. Thanks, too, to Peter, Lisa and Jack of the Printer's Press. They have worked closely with the sanctuary; indeed Lisa once went to a nearby restaurant to have a steak cooked for Raven, after he had an especially trying day of being an ambassador. This wolf preferred his meat cooked.

Thank you, too, to the staff and volunteers at Wild Spirit. They take care of wolves every day and some nights.

And to all the people who run with wolves: Thank you.

Raven with Jack photos: Leyton Cougar

Photo: Allison Bailey

photo: G Cougar

...beyond the family farm, off in the woods...

photos: Jan Ravenwolf

# INTRODUCTION

The boy awoke to night sounds, all of them ominous now and disturbing, after the dream. Usually the animal sounds were a comfort to him. He knew that he could see his gathering of friends in the morning, his pet goat and rabbits and hamsters, as he fed them, and the horses and cows and his favorite dog. But in the heavy darkness he was alarmed by the prospect of animals that he could not see, that lay beyond the family farm, off in the woods. He visited the woods often, when he could get away, after chores were done.

This is the first time that he had thought of the woods as an alarming place where anything could happen, even something bad, and not as a get-away from the rigid rules of his home. He was nine and a little shy, his mother said. His father said little, and insisted on good work.

In his troubling dream a black and gray dog had held his right arm in its jaw. The dog's teeth gripped him and hurt him. In his simplicity and kindness, the boy just said, "Please release me; please let me go."
Metaphorically speaking, that dog never let Leyton Cougar go. And the sense of mystery stayed with him, too, on his quest under the tutelage of two wise medicine men, and into the presence of fifty + wolves and wolf-dogs at the Wild Spirit Wolf Sanctuary.

This is the story of Leyton and one special wolf, an ambassador to humans, Raven.

...wolf medicine was strong medicine...
photo: Jan Ravenwolf

# CHAPTER 1

## LEYTON'S ATTRACTION TO THE WILD AND THE SPIRIT

After high school, Leyton attended a Bible School in Minot, North Dakota. To hear him tell it, he spent more time at the local zoo than in class, or absorbed in scripture. He thought that he wanted to preach. His biological grandfather had been a preacher. More to the point of this narrative, however, is the bloodline of his mother who was one quarter Lakota Sioux. Leyton, who was drawn as a young man to Indian ways, was her only blond haired, blue eyed child.

Perhaps it was his experiences as a child, immersed in nature and close to animal life, that left him impatient with the agenda of a Bible college. He took off for Los Angeles and made high end furniture and managed rock bands, to support himself, while he wrestled with his spiritual truth. He was disposed by kinship and family lore to the Native American traditions. He learned of a meeting in the Joshua Tree Desert outside of LA. A fellow seeker, whom he met at the gathering, told him of a teacher who lived just about a mile from Leyton's apartment. He insisted that Leyton find the place and that he would find what he was looking for there. It was Leyton's first sight of a "sweat lodge," standing side by side with a teepee, in the backyard of the "teacher."

The group of men whom he met that night called themselves the Seneca Wolf Clan. That initial experience led him to meet his elders, the medicine men, both of whom he called Grandpa, and with whom he travelled for six years, Grandpa Bearheart, and Grandpa Little Eagle. One of the men at the ceremony that night, perhaps in reference to the name of this particular lodge, told Leyton that wolf medicine was strong medicine. The words were an enigma at that time.

Leyton's soul search as a young man brought him to New Mexico in 1987 in the company of Grandpa Bearheart, who had been Leyton's teacher now for six years, as they travelled with medicine people healing and teaching the ways of Native American songs and ceremonies. In 1989 Bearheart settled on New Mexico as the place for Leyton to take part in a vision quest. The site of the grueling four day experience was what the courtyard of the Wild Spirit Wolf Sanctuary is now. The fire pit there is the same that Leyton used during his time of fasting and waiting for his vision. In spite of this coincidence, however, in 1989 he had no connection with what was a rudimentary beginning of a wolf sanctuary. The location simply housed a few stray dogs and wolf- dogs, under the care of two generous women. He did not meet them until later.

After the vision quest, Leyton was compelled to make a promise to Bearheart that he would pray everyday for him for a year. In return, the old man gave Leyton a pouch of medicine to wear around his neck, and his first eagle feather. After Leyton's first Sundance ceremony in 1992, the pouch fell from his neck. A sign? Who can say, but he was moving closer to what later he might characterize as his destiny.

Leyton met the two ladies who were founders of the sanctuary, as it existed, in 1994. One of them had asked him to build a geodesic dome for the property. While he was there, he conducted a sweat lodge ceremony for the young volunteers. After its conclusion, and probably still in the rush of the event, one of them undertook to break the rules and to go into the cage of a wolf-dog to feed it. Only the two women, owners of the property and caretakers of the needy animals, were supposed to do that. Leyton paid no attention to the youngster until the animal attacked the impulsive young man. Leyton flew to his aid and was mauled by the dog for his intervention. Leyton yelled for someone to come to his aid, as the wolf-dog ripped at his arm. He remembers thinking to himself, "Damn! He's ripping my flesh!" The black and gray dog repeatedly bit his right arm. All that Leyton could say was, "Please release me; please let me go." The only one who answered his cry for help was the same kid who had initiated the problem. After he and the young man backed out of the enclosure, Leyton realized that blood was dripping from his slashed arm. Subsequently, he spent eleven days in the hospital recovering from the wounds and a bout with sepsis. The physician left the wound open in order to cleanse it and prevent infection. Leyton, raging by now to get out of the hospital, told the doctor that he wanted to negotiate with her. If there was no sign of infection, would she please sew him up, "half way," and allow him to leave. She agreed reluc-

tantly that she would take him back to surgery, check out the open wound, and if all was well, close the inside of the gaping wound. When Leyton awoke, the tear all down his arm had been closed below the skin.
Upon his release, he sought out Bearheart. The old man gave him Indian medicine and told him that in two years the episode of the wolf attack would become clear.

Now liberated, Leyton jumped in his truck, placed the talisman from his vision quest and other belongings in the cab and drove home to the Bosque. During the time he was considering his future, Leyton recalls a dream in which Bearheart had asked him, "Grandson, did I give you something?" The chastened Grandson replied, in the dream that he had. "Then why is it now cast off in the belly of your truck?" Chastened Grandson immediately sprinted across the street, wearing only his underwear, to fetch the pouch. In 1996, Leyton purchased land in Candy Kitchen. He badly wanted to take part in that year's Sundance because this year it was to be performed on his land for the first time. However, he couldn't dance because his arm had not fully healed. Nevertheless, he was present. He could drum and chant. After the ceremony, he observed that the wound was completely healed.

In that same year, Leyton began volunteering for the Candy Kitchen Wolf Rescue Ranch full time, making only thirty-five dollars a week. He was to meet Raven in 1997, a little over two years after Bearheart's promise of clarity concerning his attack. A wolf had attacked him, causing a wound with irreparable scars on his body. The wolf Raven would become his soul mate and in some ways his mentor.

*Raven, the Timberwolf*

*photo: Jan Ravenwolf*

*Candy Kitchen, New Mexico*

# CHAPTER 2

## THE BEGINNING OF WILD SPIRIT WOLF SANCTUARY

The site of Leyton's vision quest and the home of a few stray dogs had become, over a short time, home of the two founders, Jacque Evans and Barbara Berge, and twenty wolf- dogs, perhaps four of them full blooded wolves. The twelve acres was now The Candy Kitchen Wolf Rescue Ranch. Jacque and Barbara, out of unadulterated generosity, had labored to take in captive bred wolves and wolf-dogs whose owners no longer wanted them, or whose uncaring humans had simply dumped them in an unpopulated area. "Candy Kitchen" got its name from some candy makers in the woods of northwest New Mexico. The candy making was a cover for an inordinate amount of sugar used in the main enterprise of these citizens, bootlegging. The nearest town of any size was Ramah. Wikipedia today describes Ramah, Tl'ohchini in Navajo, as a total area of 3.8 square miles and an elevation of 6,926 feet. In the 2000 census, there was a population of 407 souls. Ramah lies between the Zuni Indian Reservation, the Ramah Navajo Reservation, and the Cibola National Forest (Wikipedia). The town of Ramah is in a valley that has rich farmland and several huge crop fields, The terrain is entirely wooded, except for stark, red cliffs and a small man made lake.

Twenty miles from the Mormon town of Ramah lies Candy Kitchen. Its population is unknown. There are those in the woods who do not want to be found by census takers, and perhaps by anyone. The small wolf refuge was not ideally located to attract tourists, or benefactors from any place.

What is now Wild Spirit Wolf Sanctuary, the former Rescue Ranch, continues its mission, "To provide permanent, safe sanctuary for abused and abandoned captive-bred wolves and wolf-dogs. To educate the public on the wild wolf, the complexities of wolf-dog ownership and the excellent care and treatment of all animals domestic or wild."

Under Leyton's direction, the sanctuary now houses 50+ wolves and wolf-dogs in well maintained enclosures from 5,000 square feet for a pair of wolves up to one acre for a family or group. The property includes dwellings for a more permanent staff, an office, a gift shop, a kitchen for the volunteers and another for food preparation for the wolves. There are picnic tables in the office overflow area, a fountain, an arboretum, extensive landscaping, a graveled driveway, about 60 acres in all. Now there is a campground about a quarter of a mile from the enclosures. Increasingly, people come with their tents and RV's to spend the night under the stars and hear the howl of the wolves. The animals live in packs of two or three approximating their lives in the wild. Leyton utilizes a kind of dating service for pack placement. When an animal is acquired or dies, two or more animals have a trial living arrangement to see how compatible they are. The sanctuary, a non-profit, must rely exclusively on donations from generous and faithful benefactors. Without them, the work could not continue, and the wolves could not eat. It costs seven dollars a day to feed each of them.

The newsletter The Howling Reporter is compiled and edited by Georgia Cougar and mailed to ten thousand households, and seen by countless others on the web. The sanctuary has an excellent outreach program. Leyton and Raven have appeared at every major cultural celebration in the Albuquerque/ Santa Fe area, on a regular basis. They have visited thousands of school children, and been visited by hundreds more. Tours are conducted at the sanctuary daily, except Mondays, for other hundreds of visitors.

Leyton is on the road much of the time; his educational talks draw large numbers of people eager to hear about wolves, and to learn that they should not own one. He also travels to states some distance from New Mexico on rescue trips. In two recent ones he brought home two families: a litter of pups in one trip, and a whole family of a pregnant mom and pop in another. The sanctuary receives dozens of calls that they necessarily must reject. Some cases merit acceptance, but not all.

When Raven came to the sanctuary, however, in 1996, these improvements were not yet realized. In fact, one could argue that they were made possible in large part because of the partnership of Leyton and Raven and their numerous appearances.

As Leyton reflects on his early affinity for animals, he recalls his need to visit a small zoo in Minot and a very large one in Los Angeles, as often as possible, when he lived in those two cities. He would stick his arm through the space between fences and pet the wolves and the mountain lions. When he approached the zoo pens, he would whistle loudly and the wolves would begin to howl. A wolf howl is a phenomenon that goes to the core of the human psyche. Once having heard it, especially in unison with several wolves, one will never forget it.

Time after time in his life, Leyton's way bent to woods and desert, valley and mountain. He did not think much of cities. He sought the company of animals and dwelt in those spaces where they could be found. His spirituality found a form in Indian traditions. He believes that everything that occurred in his life prepared him to know the thoughts of animals, to see through their eyes. He does not call himself a wolf whisperer, nor does anyone else, but the nomenclature fits, and nowhere so much as with Raven, his brother.

Meanwhile, since his return to New Mexico and his becoming the director of the sanctuary, Leyton's life was significantly enhanced by one visitor to the sanctuary.

A young woman, a city girl as Leyton calls her, began helping out with the work at the sanctuary on weekends. Georgia Garner had a recurring dream, too. She dreamt of running with wolves on either side of her. Georgia phoned her mother (the grandmother in this wolf story), after she had spent a brief time at the sanctuary. She told her that as she walked past the wolf enclosures one night, each wolf in turn had run with her as far as its enclosure would take it, until all had accompanied her, one by one, to the end. She had run her way into Leyton's heart as well.

Grandpa Bearheart was to remain a close friend and teacher until his death in 2008. Several years after Leyton's vision quest, Bearheart performed the wedding ceremony for Leyton and his wife Georgia, and later blessed their daughter, Lakota Golden Cougar who is eight at this writing. This is a child who has honestly lived with wolves.

For Leyton, New Mexico, especially the area around Ramah, is his spiritual home and his sacred space. His family lives in a hogan which he has enlarged. It is a work in progress for sure. His skills as a wood worker are valuable in the woods in Candy Kitchen, New Mexico. Thanks to the kindness of a benefactor, they do have running water, and the electric grid extends to their house. Leyton adds to the house as time and money afford.

Leyton Jay Cougar with Raven
photo: Phil Sonier

# CHAPTER 3

## LEYTON AND RAVEN MEET

Leyton had determined to make his home in New Mexico. He set about improving his land and building a temporary shelter in which to live. He says that he deliberately stayed away from the rescue ranch at first, because he knew that he would fall in love with the wolves, and want to stay indefinitely. However, in that same year, destiny, and his financial need took over. He was broke, and he was hungry. He gave in to the habit of eating and joined the motley crew at the wolf ranch, some volunteers, the two women and George an assistant.

The ranch received a call one day, answered by a volunteer, asking if the place could accept a wolf that was about to be euthanized. The caller claimed that the animal was too beautiful, too magnificent, and at the time, too young, to be put down. The only answer that the associate of the struggling facility could give was no. The cost to fly him here was too much. Soon after, the same volunteer was giving one of the regular tours of the place and spoke about the wolf that was scheduled to be euthanized in just two days.

A tourist asked how much it would cost to get the wolf to the ranch. The tour guide gave an arbitrary figure of four hundred dollars. Tammy Cole, the visitor, wrote a check for the amount. Years later, she told Leyton that that was the last money in her checking account.

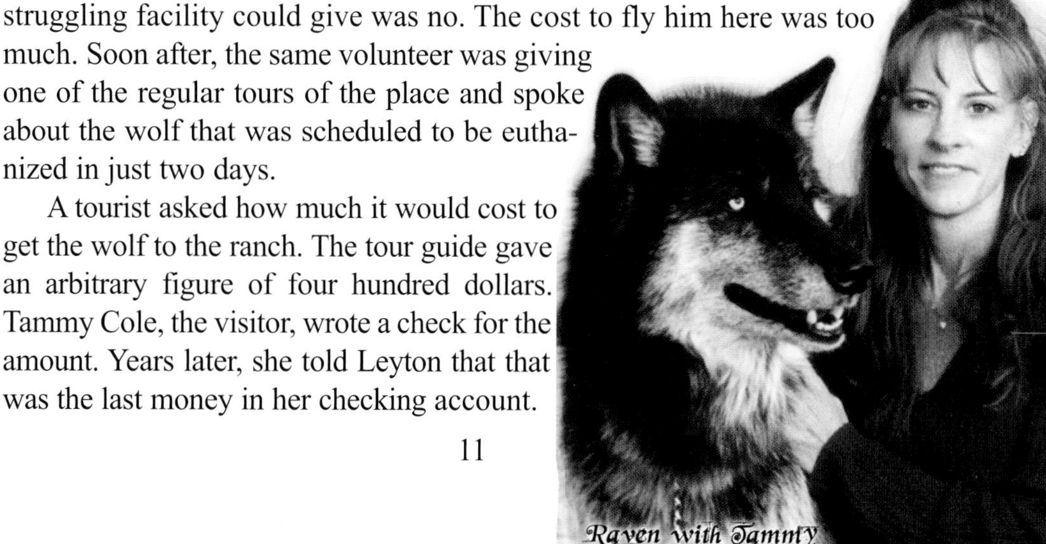

Raven with Tammy
Photo: Jan Ravenwolf

So, the magnificent creature was spared, and flown to New Mexico from California. After a mistake of routing him to Utah where he spent four hours, George picked him up in Albuquerque and brought him to his sanctuary where he would spend the next twelve years. His name was Raven.

The next day Leyton became Raven's caretaker. He says that the wolf really spooked him. Raven grumbled, snarled and growled and was as fierce as a wolf can be short of a direct assault. There is evidence that he had been abused. His background story is sketchy, but one essential element is known. He was disciplined by the owner's act of stepping on his paws when he misbehaved.

On the second day after his arrival, Raven grabbed Leyton's arm and chomped, not with his full jaw power, but enough to send a message. Leyton flashed back to the dream of his boyhood about the black and gray dog who held his arm and would not let go, in spite of Leyton's pleading. This was a black and silver timber wolf with white chest markings that grabbed Leyton's arm and held it and would not let go, despite Leyton's pleas. Leyton says now that Raven turned something on in him, something that would make communication between them easy and natural. This incident, of course, consisted in one way discourse. Leyton recalls Raven's enormous energy, with flashing yellow eyes and the stature of a champion race horse, straining and constraining at the same time, anxious to move, to run with the wind.

Raven had arrived in February; he was two years old on April 5, 1997. Later in April, on Earth Day, Jacque suggested that Leyton take Raven to an event that the ranch was to participate in and, hopefully, benefit from. She would be going along. On this first venture, Leyton and Raven were to be in each other's company longer than either had expected. The van containing Leyton, Jacque and Raven made its way to Albuquerque to attend the festival. Leyton was still very apprehensive about being in close company with a wolf for eighty miles. Things were fine until the drive home when the van had a flat.

It was midnight, very dark and deserted on the back roads off the main highway. There was no spare, no tools. Leyton left Jacque with the van and the wolf, and began to scout for some help. Jacque managed to get some sleep; she was used to being vulnerable around a wolf. Much later, Leyton returned with a tire and a jack. He decided that Raven had been cooped up long enough, so he slipped the leash over the massive head of the wolf and took him for a walk. In spite of fatigue, Leyton had a beautiful time. He talked to the wolf, that didn't grumble back at him, and seemed to form a tentative bond. That was the beginning of a long, not always smooth relationship. Raven liked humans; he liked being around the species. He was tolerant, so the way was less crooked than it might have been.

On the second trip out, Leyton soloed. He and Raven went to a Mountain Man Retreat in Tijeras. In spite of the tough costumes, the willowing coats and strange hats, Raven was fine until somebody shot a gun. At that, he took off, dragging Leyton at the other end of the leash with him. Leyton had already laid out the merchandise from the ranch, and the change box with a hundred dollars in it. Now Raven had literally made for the hills. While Leyton worried about his girl friend who had accompanied him, the money and the articles for sale, Raven wouldn't budge. He stood his ground for forty five minutes. Then Leyton discovered something that would aid him many times in the future. He had to negotiate. Everyone involved with wolves must learn this simple procedure. It is a quid pro quo game with them. You give something; you get something. So Leyton promised Raven that if he would go back to the camp with him, Leyton would never again make him do something he didn't want to do. According to Leyton, Raven looked at him and said, "OK."

As soon as they came back down from the hill, Raven made it known to Leyton that he should place several bales of hay, one on top of the other. With mountain- man help, the hay was stacked, whereupon Raven jumped to the top and remained there for the rest of the session. He could look around now, scope out his dangers, if any, and remain more content than threatened.

At this point in the story of Leyton and Raven, at the beginning of their relationship, some explanations are in order. Most readers will not easily accept the image of a wolf on a leash, or a man and a wolf "talking." The images more real to most of us are those described when Leyton was mauled by a wolf-dog, or snarled at by Raven. Wolf suggests bloody at some point in a conversation between man and Canis lupus.

The wolves spoken of in this narrative, and so well exhibited by Raven, are captive wolves. That means, generally, that they have spent some time with humans, were either born in captivity, or entered into it early in life. Zoo animals are the kind that we are most familiar with, having been in the wild and captured, or born into captivity to zoo parents. At the Wild Spirit Wolf Sanctuary, there are two families of wolves that were rescued and bottle fed by the sanctuary staff. They are now three years old. Maturity for a wolf is achieved between three and four years of age. Wolves in the wild may live to age seven on the average, but with sanctuary care to twice that. So, sanctuary animals are socialized. That human exposure is fostered by caretakers so that the animals can be fed and, when necessary, taken to the vet. The sanctuary is a rescue operation, not a breeding ground. In fact, the first order of business before a wolf or wolf-dog reaches maturity is to have it spayed or neutered.

A few months after Raven was rescued and met Leyton, Leyton had cause to leave the sanctuary. He was absent for several months. During that interim, Raven refused to eat, became quite emaciated and ill. The volunteers came to Leyton with the description of Raven's condition. He was gradually drawn back to the sanctuary by agreeing to take care of Raven, just Raven. Some disagreements were smoothed out and he returned full time. Raven recovered now that Leyton was back.

Often, when Leyton visited Raven, they would play, and play hard. The starting words from Leyton were, "Hard, Fast, Now!" Their play could be terrifying. They wrestled while making really scary deep throated growls and grunts. On a play date, one fine day, Raven leapt on Leyton in such a way that when he turned, he threw Raven on the ground and he landed on his back. The position meant that his stomach was exposed. That exposure, in wolf language, meant that the wolf was submissive and humiliated. To this day, Leyton insists that it was the wolf's fault, that he had sailed over Leyton in their wrestling. He told Raven over and over again that no one was to blame. Raven would have nothing to do with Leyton for four months. Leyton would come in to clean out the enclosure, to feed Raven, to talk with him, to tell him again and again that it wasn't his intent to defeat the wolf in such a manner. Finally exasperated, the man decided to grovel before the wolf. He got down on all fours, lowered himself to the ground as closely as possible, made whining noises, turned over to bare his stomach, all the while whimpering and whining. With those antics, Leyton made peace with one very stubborn wolf.

*...their play could be terrifying...*

photos: Jan Ravenwolf

The very presence of Raven was a lesson for school children...

# CHAPTER 4

## RAVEN AS TEACHER

In Native American lore, and that of other ancient people, the wolf is considered to be one of the primary teachers of man. In spite of remaining shy, if not actually fearful of Raven for the first two and a half years, Leyton began to accept invitations for both to appear at schools. He speaks of these events as he being the voice, and Raven the good looks. Leyton was certain by this time that he could count on Raven to be calm and restrained around children. His trust was confirmed later by a family event. When she was only ten months old, his own daughter had been left alone in Raven's presence. That was an accident. Always, before the incident, Leyton or his wife had held the baby with Raven in proximity, but she was never left alone with the wolf. Family is very important to wolves; they care for their young no matter the relation. Raven had been known to regurgitate his food for wolf pups, just as their other wolf elders might do. On this occasion, Leyton called out to his wife that he and Raven were leaving for a speaking engagement, but then he recalled that he needed to check out something in the van, and left Raven in the house. Georgia, of course, knew that the baby was sitting on the floor upstairs, but did not know that Raven was still in the house. When she returned to the baby, Raven was standing over the infant; she had a big smile on her face, and she was covered in wolf saliva. He had finally been able to get up close and personal with her and had covered her with slurpy licks. Her hair stood up like Baby Huey's.  After an initial heart attack or two, Leyton led the wolf away to his next public appearance.

Leyton, Lakota & Raven
Photo: Jan Ravenwolf

"Let's show them."

Photos: Jan Ravenwolf

This episode occurred at the home of Georgia's mother, Helen Garner. She was Raven's "Grandmother." He did try to sleep in her bed on occasion, but he never tried to wear her clothes, and he did not eat her. There will be more on his sleepovers later.

So, Leyton was fortified with the knowledge that, in all likelihood, Raven would not be aggressive with school children. Raven was always on a leash in the presence of the public, but in truth, Leyton could probably not restrain him, should he decide to take off. There is the example of the Mountain Men meeting.

Helen recalls watching Raven with an audience when a mother, so involved in what Leyton was saying, was not paying attention to her baby in a carriage. Raven's handler at the time was not paying attention to him. Raven, without being solicited, licked the baby's face and head and then backed away.

Leyton and Raven became more and more of one mind. On one occasion when visiting school children, Leyton began his explanation of ritual dominance, a practice of wolves in an effort to gain ascendancy within the pack. While speaking, he caught Raven's eye. Raven seemed to say, "Let's show them." So, the moves of domination were enacted and became part of the show for school children. The kids were terrified and thrilled at being terrified.

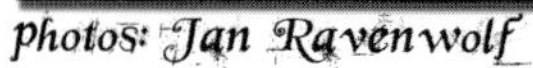

photos: Jan Ravenwolf

One of the main points that Leyton always made was that wolves are not meant to be pets. In fact, they should never be pets. The reader might be thrown off a bit here. Raven was an exceptional wolf, much less aggressive than the average animal, and much more tolerant of humans, but he was never a pet. Loved, yes, spoiled, some, but not a household pet. He lived in an enclosure at the sanctuary, not in a private home, but along side other enclosed wolves, with his mate Cheyenne. Cheyenne, by the way, earned the title bitch honestly.

Photos: Allison Bailey

Raven was the prince of the wolves at the sanctuary, but Cheyenne tried very hard to be an alpha female and Raven's boss. Cheyenne was a gift to Leyton, of a sort. He met a man who had a connection with other friends at the Sundance on Leyton's property. Leyton gave the guy a hundred lava rocks for his sweat lodge, and the new friend gave Cheyenne to Leyton. The wolf was little more than a pup then and weighed about 45 pounds. Leyton took her to Raven's enclosure; she was shaking from fear of the big wolf. Cheyenne began to try to dominate immediately, out of her fear. Leyton thought that this combination should work out, because she would keep Raven active. Leyton emphasized over and over again that no one should keep a wolf as a pet, or even as a watch wolf. The animals at Wild Spirit Wolf Sanctuary are there, for the most part, because someone tried having them as pets. When the wolf did not act like a dog, the prototype in the owner's mind, he might abuse the animal in an effort to discipline it, or get rid of it in a cruel or negligent manner. Wolves are too much to handle, not because of strength or aggression so much, as simply because of their native wolf behaviors. For example, they do not mark territory only as a dog does. They do urinate within their own territory, but they also dig trenches around their territory. So, they are inclined to dig up the family carpet.

The very presence of Raven was a lesson for school children. He stood much higher than a dog. When he walked up to a kitchen counter, for example, he could pick up a morsel or sniff around without stretching any part of himself. When he jumped to put his paws on Leyton's chest, he stood with his head higher than Leyton who is six three. He was imposing, but he smiled often and was adept at lick kissing.

On one school visit, Leyton and Raven were met at the door by an escort of students. The school was very large and consisted in many turns and stairways, a veritable maze.

After the presentation, because it was time for dismissal, Leyton did not have an escort back to the van. He had no idea of its location because he had been talking with the students on the way in, and had not paid attention. He told Raven that he should lead them back outside. Naturally, the wolf retraced their steps perfectly, and they were on their way.

Through the years the two friends have addressed thousands of school children. Raven preferred the elementary students, was all right with middle school, but was not too happy with the high school classes. He didn't like the attitude they usually presented.

When the male teenagers were just too blasé and cool acting, Leyton would often find a way to impress them. Usually the means of impression left them backing up from the wolf, or leaving the site quickly.

The wolf was abnormally patient. At another presentation, a toddler entertained herself by stepping on Raven's feet, almost stomping actually. Even though that was the same abuse that he had suffered under his previous owner, he made no move toward the child.

Raven afforded Leyton many opportunities to teach simply by his appearance and presence. On one occasion, Leyton, Georgia and Jacque stopped at a Taco Bell in Albuquerque, with Raven in the van. The three stood in line and, while waiting, an elderly couple entered; the man asked Leyton if that was his animal outside in the van. Leyton admitted to being with the wild animal and the man replied, "You know us farmers and ranchers around here don't really care for them things." The man was referring, of course, to the fear on the part of livestock owners that wolves would attack the domestic animals for food, that wolves were predators. The fears and the argument have been exacerbated in the last two decades by the wolf reintroduction program. According to Wikipedia, Wolf reintroduction involves the artificial reestablishment of a population of wolves into areas where they had been previously extirpated. Wolf reintroduction is only considered where large tracts of suitable wilderness still exist and where certain prey species are abundant enough to support a predetermined wolf population. The farmers and ranchers fear, naturally that the certain prey species will include their domestic live stock. Actually the prey species that wolves seem to prefer is elk. As Leyton says, when wolves see an elk, they don't see kill it, and blood and gore. They just see breakfast. And yes, some wolves and wolf-dogs live in places like Wild Spirit Wolf Sanctuary in order to save them from the guns of farmers and ranchers who

are protecting their livelihood. Wherever wolf reintroduction is fostered, there will be conflict.
The outcome of the dialogue between Leyton and the man in line at Taco Bell was fortuitous. The rancher/farmer was so impressed by Leyton's explanation of just what it is that he does with wolves, that the man picked up his tab and invited all to visit his ranch. Leyton's talent for public persuasion was growing, as his relationship with Raven strengthened.

In the Summer 2009 issue of "The Howling Reporter," Leyton sums up the educational part of his and Raven's programs:

*Standing before at least half a million people in our years together, we taught in classrooms, theaters, conference halls, prisons, rehab homes, churches, libraries and social halls. Raven and I have spent thousands of hours together in front of grocery stores, at art shows, at wildlife awareness events, TV and radio interviews, and would go just about anywhere people had an interest in what we do at Wild Spirit Wolf Sanctuary.*

A note about the communication between Leyton and Raven: Leyton attributes it to familiarity. They spent a lot of time in each other's company. Leyton would visit the wolf almost every day, take him for walks and visits around the sanctuary. They were on the road a lot, through New Mexico, Arizona, California and as far in the opposite direction as South Dakota. Leyton says that he observed body language, and received images. He translates these images into English language. He always spoke to Raven in English. The understanding between the two was largely a matter of proximity. One could easily anticipate the moves of the other. Of course, it sounds like there is a mystical quality operating in their exchange, but perhaps not so much. Anyone who has spent time with a beloved dog is undoubtedly amazed by how much the animal understands.

These two had an extraordinary symbiosis, true. On the other hand, humans easily endow the wolf with its historical mystique. He is prolific in our writings and our oral histories; in early American Indian mythology and in the fairy tales of the western world. He was demonized in the medieval period of European superstition and suffering. He is still a demonic figure for those who want to hunt him down and kill him, either for sport or toward protection of their land and cattle. In western civilization, man seems always to demonize what he wants to eliminate.

photo: Christine Meier

photos: Jan Ravenwolf

Kinders 2

photo: Angela Albrecht

Photo: Jan Ravenwolf

# CHAPTER 5 

## RAVEN AS HEALER

Raven's appeal to children continued in at least two extraordinary manifestations. Leyton and Raven were making an appearance at Bosque del Apache. Leyton overheard a Mom saying to a little boy: "There is a beautiful black wolf standing right in front of you." Leyton saw that the little boy was blind, just as the child said: "Oh my gosh! Wolves are my favorite animal." Naturally, Leyton wondered how he could have a favorite animal, since he was blind and had never seen any animals. Nevertheless, being the kind person that he is, he asked the boy if he would like to see a wolf as no one has ever seen a wolf before. He replied yes; so Leyton brought the boy to Raven, took his hand and put it in the wolf's mouth, petted his legs, his body, and touched Raven everywhere. They were all crying as this ten year old boy had the experience of anyone's lifetime. Raven remained patient and still throughout the boy's investigation.

Once while they were appearing at Wild Birds Unlimited in Albuquerque, a woman burst out screaming. She threw her hands up and ran out. She returned in about ten minutes with a little boy. He saw Raven and just started crying, tears rolling down his face. He asked Leyton if he could hug Raven. The boy said that he just loved wolves. So, Leyton said, "Well, let's go see one!" The grandma, as it turns out, said that she just had to go and get the boy because wolves are what he talked about all the time.

As Leyton recalled the incident of the blind boy, he was reminded of a blind man who just before his death, told a friend that he would give anything to see a wolf. When Leyton heard about the man, he said, "Let's go!" He, too, got the opportunity to see a wolf with his hands. Leyton is never surprised by these encounters. Raven knew when people needed him, and gave in to the necessity. Those responses are part of his magic. They met a lady at Whole Foods once, unknown to Leyton at the time, who was terribly frightened of big dogs because of a childhood experience. She later emailed him to say that when she got out of her car on that day, she saw Raven and she just had to approach him. The wolf and the lady seemed to lock eyes. She was wearing large sunglasses so it was difficult to be sure of that. She knelt down; he kissed her, while tears were running behind her glasses and down her cheeks. She says that she walked away cured, that she was never afraid of big dogs again. Another woman attributes her power to stop smoking, without any symptoms of withdrawal, to her meeting with Raven. In this incident, Raven would not kiss her because she had a smoker's breath. Leyton explained that to the woman, and subsequently she just quit. She never smoked again.

Another time in the early days, a woman came to the wolf sanctuary and told Leyton that she had cancer and that something told her that she might be healed if she could touch a wolf. At the time, Leyton wished that she had not said, "might." In spite of her tentative demeanor, he answered, "Then let's go touch a wolf." he took her into Raven's enclosure, and gave her something from the medicine pouch, as well as a claw that had come from one of the wolves. After she left, she wrote a letter and enclosed a check. She sent a check every month thereafter until she died. When Lakota was born, the woman sent a baby blanket that was a gift to her from her mother, and before that, from her grandmother to her mother. Now it reposed with Leyton's daughter.

photos: Jan Ravenwolf

Raven simply brought a thrill wherever he went and to whomever he touched. Everyone believed that it was a privilege to meet him. No one had expected to meet and be happily licked by a wolf, or to sink his fingers into the two layered, thick, deep fur. Leyton says that Raven was a connection between wolves and humans. He

was exceedingly generous to let himself be pawed over by countless children and adults at any one appearance. Leyton recalls that he only snapped once, and that was at a fly. Apparently wolves don't like flies, but then who does?

Raven spread joy to all who met him. He would stand for hours at the end of a loosely held leash while Leyton explained the place of wolves in the ecosystem, the mating and denning habits of wolves, their behaviors in the wild and in captivity, the work of the sanctuary, whatever anyone in an audience or at an exhibit asked. Too bad for Raven-Leyton loves to talk with people, and does it well. Raven would kiss (lip lick) anyone brave enough to bend down to him. Eventually, he got to be everyone's surrogate pet wolf. No, wolves don't make good pets, but this wolf was Raven! Sometimes he brought joy by just being himself, another creature outside of our realm of reference.

A story that gets told about Raven was the time that Leyton went into the wolf's enclosure to find him a little nervous, trying to keep Leyton away from a certain spot under a tree. Well, that just made Leyton all the more curious, so he found the place where Raven had buried some of his food, trying to hide it from Cheyenne. A little later, Leyton took Raven on a road trip; when they returned, the food was gone. Cheyenne had raided his larder, so Raven was forced to locate another, better spot. On a cold frozen day Leyton discovered his secret place. This time a sheet of ice covered the hole, so he began to refer to it as Raven's refrigerator.

Raven was a gentleman. When I first became acquainted with the family, I was seated in a back room in the home of Helen Garner, Raven's grandmother, a.k.a. Georgia's mother. Leyton let the wolf in the front door without any announcement or greeting, hoping to get a reaction out of me, I guess. Raven loped to the back room; before I knew what was up and before Helen could warn me, he put his head over the three foot high couch back, and licked my cheek. I was officially in. Now that I think of it, Leyton and Raven had perhaps discussed the move before entering the house. Maybe it was Raven's idea.

After the death of his wife resulting from cancer, Pablo Rosales wrote the following for publication in the Fall 2009 issue of The Howling Reporter: (his words are far better than mine).

*Twelve years ago, my wife Debra was diagnosed and treated for cancer, one which allowed for only a 50% survival rate past five years. Although we live in California, we began to visit El Santuario de Chimayo to give thanks for all the blessings in our lives and to seek the strength for both of us to be able to handle and endure whatever the future would bring us. Because of a very spiritual experience that we had during that time, we made our first visit to the Wild Spirit Wolf Sanctuary as we discovered that Debra's guide through her journey was the wolf. At our first visit to the Sanctuary, my wife and I got to meet several of the wolves during our tour, but it was Raven that seemed to make an impression and impact on her. She was in awe of his beauty and strength, yet informed me that he also had some pain in his heart but would not show it to others because his role was to give. . .this before we knew anything about Raven. She asked me if I thought she was being strange (because she felt strange at that moment). . .What I knew was that she was being given a once in a lifetime gift. She would experience a special connection to Raven as she looked at him, and he appeared to do the same. Describing this is not easy; many may quickly discount its validity, but for us, the feelings that were felt at that moment were those of a connection that remained, more so for Debra, forever.*

photo: Angela Albrecht

Jan Ravenwolf is an important person in the life of the Wolf Sanctuary. She knew Jacque and Barbara before she was acquainted with Leyton or Raven. She has served as a board member for years, and as the photographer for most of the sanctuary's publications. She is the official and very generous photographer for this manuscript. I tell her that we would have not even half a book, and a much less evocative one, without her photos. Jan was kind enough to mention her own past abuse, always a difficult topic, in the context of Raven's unconditional acceptance. She says that she had "hidden in plain sight" for most of her life, with little self esteem or feelings of worthiness. By the way, Jan had taken the name Ravenwolf before meeting the animal by that name. Prophetic? There was some kind of foreshadowing that is hard to dismiss. Regardless, Raven and Jan had a special relationship that heartened her, and allowed her to feel a peace that she had not known previously.

photo: Jan Rogers

There is a lot of discussion in our current culture about animals' capacity to communicate-with each other and with the human species. There are now animal psychologists, and animal trainers who can read the behaviors of domestic animals, and there are known to be animals that can imitate or originate abilities that seem extraordinary. Before our era, we credited animals only with instinct. We were careful, especially in our theology, not to endow them with a soul. St. Francis of Assisi, in his hagiography, was allowed to talk with all forms of life. Those communications were considered miraculous, however, therefore exceptional. The corollaries, if we allow animals to have spirits, are astounding. Can we then still kill them for food, for clothing, in order to control the environment, for the sport of hunting? What does it mean if we admit that animals can think? We seem to see our pet dogs and cats thinking all of the time. Is the mind separate from the soul? Henry Beston, a naturalist from the early 20th century, says in The Outermost House, "In a world older and more complete than ours they [animals] move finished and complete, gifted with extensions of the senses we have lost or never attained, living by voices we shall never hear. They are not brethren, they are not underlings; they are other nations, caught with ourselves in the net of life and time, fellow prisoners of the splendor and travail of the earth." Barry Holstun Lopez in his book Of Wolves and Men, quotes Beston, and then ends his eloquent book with this conclusion: "To allow mystery, which is to say to yourself, 'There could be more, there could be things we don't understand,' is not to damn knowledge. It is to take a wider view. It is to permit yourself an extraordinary freedom: someone else does not have to be wrong in order that you may be right."

Still, if we consider the human species' treatment of those judged to be inferior, inferior humans, that is, it would appear that we are a very long way from affording animals their due as possessors of body and soul, language and morality. In the course of human history, wolves have been represented as werewolves, as metaphors for marauding and voracious men, as consorts to witches and as the devil himself. The hour of the wolf is the danger zone just before dawn when man may be drawn into the strange and lascivious. A licentious man is referred to as a wolf; a deceptive person as a wolf in sheep's clothing. We seem to identify our "lower nature," our corporeal qualities with wolves, quite handily. But wolves can be shy and easily startled. The example of the gun at the mountain men camp is telling. Raven was also afraid of ceiling fans, PA systems and the air filled figures that dance around to attract attention to a sale, or other retail events, as well as skis carried over the shoulder. Leyton began wearing a cowboy hat occasionally in order to desensitize Raven to his fear of hats! I observed him once at an outside festival at Las Golondrinas, an annual event for the Spanish descendants of the state. After a cannon had been shot, he jumped up into the back of the van and would not get down on the ground again, no matter the amount of coaxing by Leyton. He would still greet visitors, and spread lick kisses to those who approached him, but he stayed in the van for the remainder of the day.

The most charming reaction of his wolves that Leyton describes is their ability to sense fear, on the part of the humans sometimes, when they are in contact with wolves. "The humans are afraid! What are they afraid of? Then I am afraid, too." The animal does not realize that the fear he senses in the humans is fear of him.

photo: Georgia Cougar

photo: Jan Ravenwolf

# CHAPTER 6

## RAVEN'S SLEEPOVERS

Candy Kitchen, New Mexico is at least a two hour drive from Albuquerque, the largest city in New Mexico, and the location of many activities that are fertile opportunities for Leyton and Raven to educate the public about wolves in general, and their home, Wild Spirit Wolf Sanctuary, in particular. Consequently, Leyton and Raven had cause to spend many nights in the city. It was always too late to drive home after an exhausting all day event. Leyton was hoarse from talking and Raven's tongue was tired from kissing so many people. Georgia and Lakota often spent the night, while Leyton was on one of his educational trips, with her mother in the big city. It was convenient for Leyton and Raven to join them in Helen Garner's home for a sleepover occasionally. In fact, Raven had his own room in the house. And other habits, almost rituals, became ordinary.

Helen has a tri-level house.

photos: Jan Ravenwolf

Raven's room was upstairs from the ground floor, while the Cougars slept one floor below. Actually, Raven's room was Georgia's old room in the family residence. When Raven began his nightly trysts, Helen had a dog named Denny, a boxer. Denny loved Raven and trailed him mercilessly every where the wolf went, except to bed. When the dog would put his paw on the first step, Raven growled. Denny slunk away. At other times, he was all over Raven, so that Helen was compelled to say: "Raven, if you kill Denny, I will be very sad, but I will understand."

Raven had a lot of dog friends. Once, really by happenstance, Raven attended Yappy Hour at Three Dog Bakery in Albuquerque. Every Wednesday Kim Snitker, the owner, hosted a time for dog lovers to bring their pets into the store to socialize and get specials on baked goods. Georgia recounts, in The Howling Reporter, that Raven had a great time. Georgia and Leyton had stopped just across from the bakery to get a cold drink when they made the acquaintance of this enterprising lady and her hilarious store. Georgia says, "Raven was a little shy; he crept up to a few dogs very submissively, but by the end of the evening, he was rushing up to new dogs as they arrived, eager to sniff his next new pal."

At Grandma's, Raven had the run of the house. He was shown how to use the doggy door once and that was enough. Raven preferred his meat cooked and given to him by a female. That worked out well for "Grandma," who cooked pork loin for him. No, that testimony should be reversed. That worked out well for Raven. The room he had chosen for himself was carpeted, but occasionally he jumped in a bed. When he did, no one tried to move him. He was in the bed for as long as he wanted. Raven rarely challenged anyone, but when it came to a soft bed, he wasn't moving. More than a few nights, someone in the family was left wondering, now, where do I sleep? Usually he returned to his own room before bed time. On his first visit to Grandma's house, he crept into her room after lights out. She awoke to see, barely, this dark form sailing over her. She leapt from the bed as Raven jumped into it. He always went into Helen's bathroom, climbed into the tub, took her washcloth in his mouth, brought it in to the bedroom and dropped it on the floor. Then he began to wash his face with it, by rubbing his face and neck against the dropped cloth. He performed this visiting ritual on the very last visit to Grandma's house when he was just short of disabled.

Helen Hugs Raven
Photo: Jan Ravenwolf

photos: Helen Garner

Photo: Liz Parr

photos: Georgia Cougar

He was known to bury a few necessities against hard times in her backyard, too, once a whole package of English muffins, another time, Lakota's Fisherman Bear. The bear was a favorite stuffed animal of Lakota's. Georgia and her mother speculated about its location for weeks. One day, Helen spotted a cone shaped pile of dirt in her backyard. On a closer look, she saw one tiny upside down foot sticking out. Fisherman Bear had been neatly buried, upside down in a deep hole. Wolves don't bury just anywhere as dogs do. They bury against a boundary, like a backyard fence. The package of English muffins turned up nearby. Georgia wrote about the lost bear for the reporter, "I'd never noticed Raven eyeing Fisherman Bear. Had he been planning the bear-napping for awhile? Exactly why did he want him? He'd been back to my mom's house two times since the burial. Did he go check on his prize, play with him and re-bury him? What was he saving him for?" Georgia assures us that with a little counseling, the bear will recover completely.

Regular contributors to the Sanctuary, Shirl and Joe Griffenberg, were so amused by the story of Raven and Fisherman Bear, that they sent Raven a teddy to call his own. The regal wolf carried Griffenberg Bear around for days, and fiercely defended his turquoise toy against any who tried to cuddle with it. It came as no surprise when Cheyenne eventually shredded Griffenberg Bear into oblivion, but the photos of the wolf and his beloved stuffed animal remain among the favorites of the visitors to the Sanctuary.

Helen recounts an early visit of Raven's. She was still a little timid. He was sprawled in the kitchen early one morning. When Raven sprawled crosswise, he was as long as the kitchen, with its appliances, was wide. Helen was going to have to step over Raven in order to leave the house for work, and she had to gather her things before leaving. She knew, from her many experiences that one does not just step over a sleeping animal. The animal is threatened by the action. Finally, she just talked to him, explaining that she had to leave for school, and took one giant step. He looked at her; his huge head turned to watch her from his bright amber eyes, but showed no other reaction. She walked back and forth for a few minutes without incident, and was soon out of the door, intact.

Later, Raven went for a block of cheese in the refrigerator. Both Helen and Leyton, at this time, went at Raven from two different directions, trying to distract him by offering other delicacies. Leyton was concerned that they were on their way to a day long show and tell and that the whole block of cheese was not a good beginning for the wolf's digestive system. Neither one of them remembers how they got the cheese away from him, but they did. Another plight that one hears of in stories about captive wolves or wolf-dogs as pets involves the refrigerator. These intelligent beings watch the owners putting in and taking out food from the big box all of the time. When opportunity strikes, they can easily overturn the fridge and feast away. And make a mess. One rarely hears of a pet dog engaging in this endeavor well, maybe trying.

top and left photos: Jan Ravenwolf

Ann Wallace was another good friend of the sanctuary and of Raven's. In the Winter 2002 issue of the newsletter, then titled The Candy Wolf, Ann's living room is described as a total wolf experience: paintings, stuffed toys and a multitude of tiny sculptures on tables. Leyton and Raven visited her home one Christmas season. She had a spread of goodies and her usual display of knick knacks. Leyton must have led Raven in with some hesitation. Displaying his good manners, Raven's tail stayed under control, while he deftly picked out one walnut from a dish, put it on the floor to crack and eat, and then returned for a second. Leyton was so proud! On one visit, however, there was a repeat of the cheese theft, right out of Ann's refrigerator. Again, the cheese chase was on, and again Leyton managed to snatch the cheese from the wolf's mouth. Ann had a houseful of ferrets, too, but Raven totally ignored them. Apparently they weren't on the menu.

There were times that didn't go so gingerly. Before Lakota was born, Georgia went with Leyton and Raven on one of their road trips. This time they were to stay overnight in the guest house of friends. Leyton got up in the night to go to the bathroom. When he returned, Raven was standing in the bed over Georgia, who, new to the Raven and Leyton relationship, was terrified. Raven seemed to be laying claim to Leyton's mate and growling all the while. Leyton tried every trick, every effort at cajoling the wolf to move away from Georgia. Nothing worked. At his wit's end, Leyton picked up the leash and told Raven that he was going for a walk. When Raven followed Leyton out of the room, Leyton quickly doubled back and closed the door behind him. The tricked wolf snorted at the door for the rest of that night. Georgia refers to this episode as, "The Battle Over the Broad."

Raven has slept in motels with Leyton on several road trips. It was always understood that Raven slept on the floor and Leyton in the bed. After a particularly grueling day in Vail, Colorado, in a crowded and busy venue, Raven entered the motel room, as stressed and exhausted as Leyton, and jumped onto the bed. Again, no matter what the attempts on the part of Leyton, Raven was not budging. The old joke about where does the wolf sleep applies; the answer, anywhere he wants to. Raven slept on the bed that night. Leyton slept on the floor.

In their early appearances together, especially when they journeyed to Bosque del Apache, it was always an overnight stay, Leyton had a camper truck. On one winter night, the temperature fell much more than anticipated, and Leyton was caught with two thin blankets and no heat. As he explains, Raven was not cuddly. Well, what does one expect from a wolf? On this night, Leyton begged Raven to join him in his cold bed and help out with some body warmth. He pleaded and explained to no end. Finally, about four in the morning, after some more cajoling, Raven leaped up to the bed and joined Leyton with his back to him, side by side, until morning. Raven liked soft beds, but preferred no one else in them.

One of their more memorable trips took them to California. They attended the LA TV Environmental Media Awards, not as recipients, however. The trip was long and tiring. Leyton drove around until he could find a restaurant that had an outdoor venue. He knew that Raven was at least as hungry as he was. He spotted an Italian restaurant with al fresco dining, and ordered lasagna. Raven drew a crowd as he always did. Leyton began conversing with the crowd as he always did. The waiter brought the order, and before Leyton could turn around, Raven ate the lasagna. The crowd found this exercise much more entertaining than Leyton had, but one kind man gave his order to Leyton and said that he would order another. Raven seemed to inspire generosity wherever he appeared.

Photo: Jan Ravenwolf

Leyton attributes their unusual rapport and singularity of minds to their travels together. They were co-dependent at these times. Jan, who frequently met up with Raven and Leyton at their outreach programs, remembers one time when Leyton walked away from the booth to take a break. Raven plastered himself to her leg and did not take his eyes off of Leyton's direction until he saw him again.

Another trip took them to South Dakota to a ranch on the Cheyenne River. The event was called Artists Ride. The concept was that photographers and visual artists could take pictures of cowboys, horses, Indians, herds of animals, and other "wild" things, including Leyton and Raven. The artists took photographs in order to take them back home where they might paint them from the stills. There were appropriate backgrounds meant to imitate the old West. Jacque was with them again. At this time Leyton would have said that Raven had no need for him, and that his need for Raven consisted in his ability to be an ambassador for the sanctuary. Raven and Leyton were together at this venue all day and night for a week. Leyton had the wolf on a 25 foot lead that was tied to his wrist. They slept that way. At one point, Leyton just had to get away for a bit. Jacque was present to testify that Raven was genuinely concerned at Leyton's absence. The wolf whined and paced until his return. There would come to be an intense trust between the two, so much so that they could play out the ritual domination that we have written about earlier, wherein a wolf allowed a man to put his head over his shoulders, without taking off that head!

There are the funny episodes in the middle of their adventures also. Leyton loves to tell about the time when he was at a gas station to fill up and to give Raven an exercise break. He had walked the wolf around the pumps and outside until the fill-up was completed, and then took Raven inside while he paid. There was a group of men around a card table who stood up and exited immediately. The clerk asked Leyton not to bring the dog closer because he was afraid of dogs. Leyton replied that he was not to worry, "that's okay; it's not a dog; it's a wolf," whereupon the man ran to the back of the building. A woman finally came out to accept the payment. I think that Leyton has pulled this on more than one occasion.

Cheyenne River, South Dakota

photo: Phil Sonier

photo: Jan Ravenwolf

# CHAPTER 7

## THE BIG BAD WOLF

What of wolves as predators? Predation is the quality that the public associates most with wolves. Little Red Riding Hood and The Three Little Pigs are many a child's first impression of a wolf. People who have to share territory with wolves are happy to keep those folk tales alive. Leyton explains the issue in this anecdote:

I took a walk in the woods with Rhino, my dog, and Raven. Rhino chased bunnies. Raven doesn't know bunnies; he didn't chase them. Once, before we married, Raven and I were visiting Georgia in the city. There was a bunny in the parking lot; it flattened down on the pavement in order to be invisible, so I didn't see it. We walked past it, but then I caught it out of the corner of my eye and said to Raven, "Hey, look. There's a bunny." He looked at it and kept walking. Another day, I was walking in the woods with Rhino and Raven again. Raven was on the leash; Rhino saw a bunny and started chasing it. Raven looked and asked, "we chase those?" I answered in the affirmative. He took four steps toward it and stopped.

On a motor trip, at a Colorado rest stop, there must have been over a hundred bunnies playing in the field next to the stop and around it. They ran through Raven's legs, all around him, almost over him. He did not blink. Leyton had said that they were going for a run, and that is what Raven intended to do. Only that. Leyton says that, on another woods walk, I saw some cows and said to Raven, "Let's go chase some cows". So we began to chase them, but there were some calves in the group. The herd gathered

Photo: Jan Ravenwolf

around the calves and charged us. Raven said, "Let's don't chase cows; cows are scary." I took him over to a cattle truck that we came across in a parking lot. He wasn't at all interested. He wanted to smell them, but he wasn't going near.

Wolves imprint on their very first food. If cows are not part of their early meals, they can walk through a field of cows, or meet a buffalo nose to nose, and never try to make a meal out of them. If mom or dad never brought a bunny home for food, then bunnies are not on their menu, until they discover that, if they chase a bunny and kill it, it is food. Or they watch another animal kill and eat, then it is a discovery and after that, on the menu. So, their predation appears to be a learned behavior as well as an instinct. Wolves and wolf-dogs, that are born into captivity and never taught an existence in the wild from a parent, cannot just be dumped in the wild. They will not have survival skills. Sanctuary animals cannot just be turned loose. They, also, cannot be turned into pets. Leyton, with the help of the transportation department, has been able to provide road kill to the wolves. In his role as Dad, he has initiated the young wolves to the diet of elk and other animals.

Leyton gets very angry about this example of one woman's naïve attitude toward owning a wolf. She called the sanctuary one day. Leyton happened to pick up the phone. She said, without an introduction or prelims, "I think I need a wolf." When Leyton asked, why? She replied that she knew that wolves were very spiritual and she needed that around her. As Leyton said, she hoped for spirituality by association. He hung up on her. He said, had he talked further, he would have raged through the phone. It would not have been nice.

Leyton reminds his audiences that man is the biggest predator of wolves, and that it takes immense courage for any of his wolves to stand before a group of people and let them come close and touch him. In addition to everything else that Leyton and Raven did in educating the public and taking care of captive wolves, he admits to working to dispel the myth of the big bad wolf.

On another occasion in the Bosque at the Festival of the Cranes, Raven saw his nemesis, the Blue Goose. This character is about seven feet tall, a costumed man of course, with large flapping wings. He is this festival's mascot. Raven is really afraid of this guy. He reacted to him as he always did; he began backing up. A boy who witnessed the encounter yelled, "He [Raven] is going to kill that goose!' Leyton gets tired of this kind of projected response. In fact, Raven was trying to get away from the huge goose spectacle.

There are those who are interested in owning a wolf or a wolf-dog because they appear as big and bad. They are perhaps the extreme of those who own and breed dogs for pit fighting. Wolf-dogs are not bred to be family pets. Their aggression is often cultivated, nevertheless. Instead of complementing spirituality, like the caller above, they are meant to complement an impressive physicality, somehow to add by association an intimidating presence for the owner. Ironically, perhaps, the wolf or wolf-dog may turn his learned aggression against the owner, or member of his family, become destructive and then be cruelly punished, and if lucky, end up at a sanctuary. The alternative is too sad to contemplate.

All humans tend to project their own strengths and weaknesses onto inanimate objects and other living things. If this were not so, we would miss a lot of comedy in our lives, and perhaps a lot of drama. There is incongruity in endowing animals with the antics of humans, therefore there exists a field for comedy. We are much quicker at performing projection than we are at being receptive to the real state of an animal's communication. Most of us miss the cues. Lately, in our culture, we have met a dog whisperer and a horse whisperer, but we don't know of a horse or dog listener. Maybe we have one in Leyton Cougar who listens to his animals. Couple that feature with Raven's determination to speak to humans and some amazing things have happened. For one, Raven listens to humans. How else could he bring the peace that he did to his human encounters, like the day he met Leyton's actual grandparents who remained uneaten.

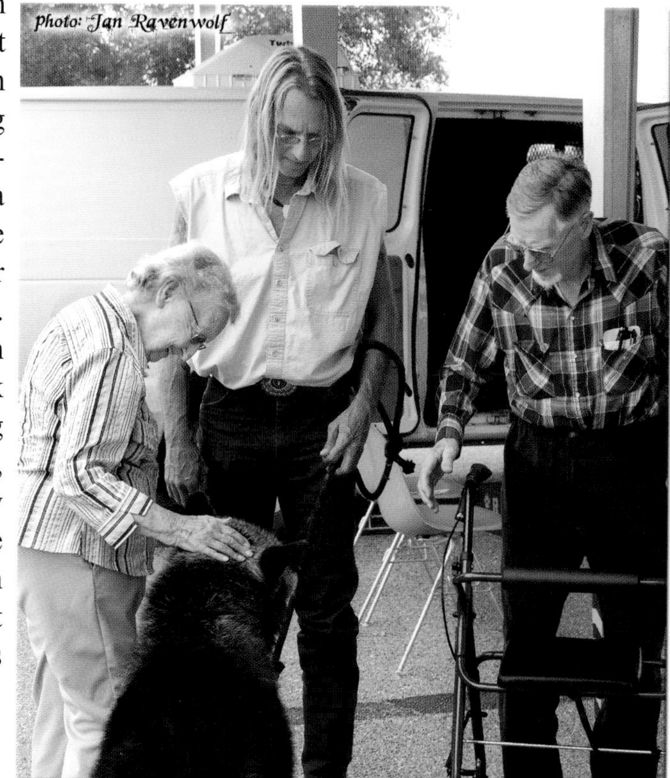

Leyton's Grandma Edith & Grandpa William

Brutus, The Wolf

photos: Angel Bennett

Choctaw, The Wolf-Dog

# CHAPTER 8

## WOLVES AND WOLF-DOGS

Throughout this book, reference is made to wolves and wolf-dogs. The latter term may not be familiar to most readers. The easiest reference would be to a half and half concept, i.e., one parent was a wolf and the other a dog. True, dogs and wolves can breed together. However, in any one animal, the DNA imprint may come from several ancestors and result in a varied content of wolf and dog. Angel Bennett, Education and Administrative Assistant at Wild Spirit took the photos and worked up some identifying characteristics for the following article in the Winter 2008 issue of The Howling Reporter.

## WHAT'S HIGH-CONTENT ?

While proof-reading our last issue of *The Howling Reporter*, our dear friend Dr Elizabeth Parr put down her copy and asked, "What does 'high-content' mean?" It's easy for those of us who use terms like this everyday to forget that not everybody knows what phrases from the wolf-dog world like "content-level" mean. We hope the following article will help answer some of your questions.

We all know already that a wolf-dog is a canine who is part wolf and part dog. Past this simple fact, things can get really confusing. What needs to be clear is that wolf-dogs aren't simply 50% wolf and 50% dog. So, how can anyone tell how much wolf is in a wolf-dog? Currently, there is no DNA test that can break down the number of wolf genes vs. dog genes, in any given wolf-dog, to give us an accurate percentage. Wolves and dogs are nearly alike genetically with roughly 2/10 of one percent difference between them.

Without any truly scientific way to determine how much wolf is in a wolf-dog, we rely on what we know: what wolves act like, and what wolves look like. The phrase "content-level," is used to categorize wolf-dogs only. We use three levels: high-content, mid-content and low-content. We place each wolf-dog into a category based on two things. How much does the wolf-dog look like a wolf and how much does the wolf-dog behave like a wolf. If he's mostly wolf, we call him "high-content." If he's mostly dog, we call him "low-content."

When deciding content level, we take into account our own experiences and many other more subtle differences. Let's make it easy and pretend that there are 7 physical characteristics and 5 behavioral characteristics that are different in wolves and dogs which we use to determine content. If a wolf-dog has more than 6 of these characteristics combined, she is considered high-content. If she has 5 or 6 of these characteristics, she is mid-content.

We'll use Wild Spirit resident, Rain, as our model. Rain is a high-content wolf-dog. She is mostly wolf in her appearance and behavior, but she obviously has a little dog in her too. We'll start with our list of physical characteristics and show you how Rain scores.

1. EARS

Wolves, like Doc at the right, have rounded ears that are more proportionate to their head size. Thick fur grows inside the ear so that just by looking, no skin can be seen inside.

Rain's ears are large and they come to a sharp point on top. Her skin can be easily seen inside of her ear. Rain scores: 1 dog point.

2. EYES

Wolves' eyes are at an angle like Ghengis' and are almost always amber in color, but rarely can be gray or green. An adult pure wolf's eyes are never brown or blue. However, a wolf pup's eyes are blue until they are several weeks old. Rain's eyes are slanted and amber in color.

Rain scores: 1 wolf point.

3. & 4. TAIL

Wolves' tails are long, bushy and always hang straight down like Forest's. Their tails are never curled. Wolves always have a spot near the base of their tails which marks their scent gland. Rain has a long busy tail with a black spot, but it is also curled.

Rain scores: 1 wolf point, 1 dog point.

5. CHEST & LEGS

Wolves' chests are narrow like Powder's. You can fit 3 fingers in between where the top of the front legs meet. Their front legs are long and lanky. The back legs, however, are cow-hocked outward, so that when a wolf runs, his front legs can come easily between his back legs. Rain has a narrow chest, lanky front legs and cow-hocked back legs. Rain scores: 1 wolf point.

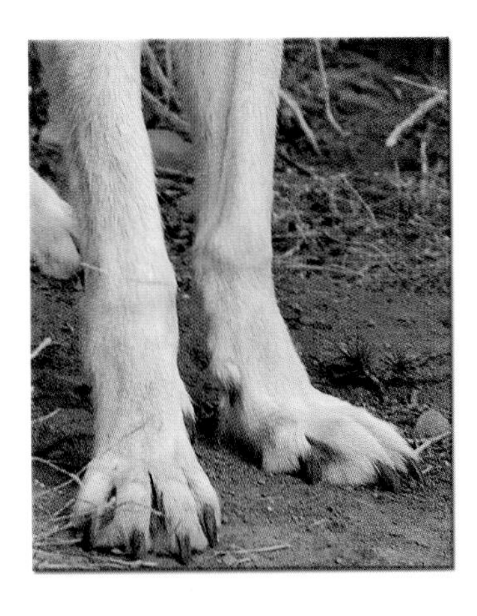

6. FEET
Wolves have large feet and elongated, more finger-like toes like Nayati's. Their toes also have webbing between them. This helps them spread their toes apart and use their feet like built-in snow shoes when going over snow drifts.
Rain's feet are disproportionately large, with elongated and webbed toes.
Rain scores: 1 wolf point.

7. FUR & MARKINGS
Wolves have thick, straight fur with a grey undercoat like Taza's. They have a section of longer fur across their shoulder blades called a "cape." The cape fur can be up to 6 inches long, and is often marked with black tips.
Rain has typical wolf markings, and a cape of hair around her shoulders
Rain scores: 1 wolf point.

Rain's total physical score: 5 out of 7 of the wolf characteristics, written 5/7.

Now, let's score Rain on her behaviors.

1. NEOPHOBIA
Wolves are neophobic which means they have a great fear of new things. For example, during our recent Howl-A-Ween fund-raiser, most of our residents eagerly awaited their familiar stuffed pumpkins and quickly devoured the contents. By contrast, wolves Akela, Brutus and Navar of the Iowa trio were frightened by their first experience with their pumpkins.
Rain is afraid of new things,
so scores: 1 wolf point.

2. DOMINANCE DISPLAYS
Dominance displays toward other canines, and unfortunately toward peo-

ple, are a part of a wolf's every day life. Those displays can include mounting, mouthing, growling and body postures.

Rain is always testing her pen-mate, Nimoy, to see who's boss

Rain scores: 1 wolf point.

3. TAIL LANGUAGE

Wolves raise their tails up as a sign of aggression. When a wolf hangs its tail straight down and wags just the tip, they are sending the message that they are happy to meet you. By contrast, when a wolf's tail is out and wagging back and forth, they are sending a warning that they are happy to eat you. Dogs often raise their tails to show happiness or excitement. We are all familiar with the happy wag of a dog's tail, which usually means a dog is happy, friendly and safe to approach.

Rain uses a dog's tail language, so scores: 1 dog point.

4. VOCALIZATIONS

Wolves howl, growl, whimper, whine, and huff, plus all kinds of other vocalizations. However, wolves do not bark.

Rain does not bark so, scores: 1 wolf point.

5. ANGLE OF APPROACH

Wolves usually approach from the side. Wolves only approach head on when they are being dominant, extremely confident, or aggressive. My labrador, Brody, will come running straight at me when he's happy and submissive. Rain approaches happily and submissively at an angle.

Rain scores: 1 wolf point.

Rain's total behavior score: 4 out of 5 wolf behaviors or 4/5.

Rains's combined total: 9/12 = high-content.

The unfortunate thing about using content-levels is that they are debatable. While many respect the opinion of folks with experience in the wolf-dog world, there is no blood-test we can give Rain to prove that she is mostly wolf with a little dog.

For practicality's sake, however, content-level is the best description possible. After all, when someone calls us saying they have a 98% wolf-dog who needs a new home, it doesn't automatically tell us that the wolf-dog belongs with us. We've found most people think their wolf-dog is more "wolfy" than we do. If we see pictures, hear about the wolf-dog's behavior and it sounds very "wolfy", chances are she can't go to any domestic home and needs to come to a sanctuary like ours. However, if the 98% wolf-dog is friendly, good with kids, walks great on a leash and loves to be around people, she certainly doesn't belong in a wolf sanctuary. Instead, she needs to be treated like a dog, and find a new family to live with. All too often, percentages end up "black-listing" dogs who could live happily in a normal domestic life.

..............................................................

Obviously, the above is not scientific. These characteristics might prove helpful if someone is experiencing difficulty with a domestic animal and trying to decide what its future holds. Some people need bragging rights and want their "pet" to be as wolf like as possible. Many people, who think they own a wolf, actually don't. Owning a wild animal is against the law, but it is also a risk to the owner and family, as well as others who come in contact with the breed. Grandma Helen wants to add, after seeing Raven's paw prints on her bedding, that his paws were as large as a salad plate.

Leyton likes to say that the domestic and the wild don't mix. Nowhere is this saying more evident than in the wolf-dog. Common knowledge among those who have one, or know one, is that the wolf-dog has inherited the negative traits of both. If one is looking for a ferocious watch dog, it might do, but usually not as a family pet. However, there is a new organization of people who would like to see the wolf-dog emerge from its negative history under the care and guidance of people who know how to handle it. Since there will always be wolf-dogs, that kind of organization is welcome.

Leyton almost lost a leg to an angry wolf-dog named Nikki. His family came to visit him. I am almost tempted to say "in jail." He was too aggressive for their home and so was rescued. His home now at WSWS might, rather than jail, be compared to a resort, but without the soft couches and carpets found in a home. Nikki was very happy to see his former family. One of them asked about a toy that was inside his enclosure. Leyton offered to get it. Nikki attacked Leyton with every tooth and claw in his arsenal. It was only later that Leyton thought that it was really his fault. To Nikki, now looking at his primary family through a fence, he saw Leyton as the bad guy. He was keeping Nikki inside; he couldn't get to his family on the outside. As soon as Nikki attacked Leyton, he jumped atop the dog house that was a part of his enclosure. The dog house collapsed under his weight. His flesh was once more ripped and torn.

Nikki

Photo: Allison Bailey

It was the boy in the family that came to his rescue. Leyton was taken to the little Zuni Hospital where his wounds were irrigated and sewn up. The dog had bitten him at least twice. The tear in his leg exposed the bone. Leyton made no move to execute the dog, or change anything about its life. He said that Nikki was just doing what might have been expected. Leyton rarely assigns blame to anyone of the canine persuasion.

Photo: Georgia Cougar

It is never hard for humans to romanticize a wild being that looks like a domestic dog. In the charming first novel by Dorothy Hearst, Promise of the Wolves, the heroine of the story is a pup who comes from questionable origins. Her mother is from the wolf pack, but her father's blood is unknown. There is the slightest suggestion that it could just be a wolf-dog, already evolved 14,000 years ago. To add to the little wolf's trauma, she likes humans. At the end, she agrees to a pact with the Greatwolves whereby wolves and humans of the Wide Valley will hunt and learn together and form a pack. A wolf that in life had followed much the same path watches her from the spirit world and feels a weight lifted from her heart. The old promise of the wolf is never to consort with humans; never kill a human unprovoked, never allow a mixed-blood wolf to live. The renewed pact has wolves and humans hunting together peacefully, and the little wolf with mixed-blood will lead them.

*photo: Georgia Cou*

# CHAPTER 9

## WHAT'S IN A NAME ?

Leyton surmises that Raven's name was chosen for his color. He was black, except for his chest markings and the sprinkling of silver throughout his body. There is, however, a symbiotic relationship between wolves and ravens that is worth remarking on here. The raven is a scavenger of food of every kind. They will follow the wolf in order to eat his leavings from his kill; wolves will follow circling ravens in the hope of finding food below. Biologists have remarked on the playfulness that seems to exist between wolves and ravens. Lois Crisler believes that wolves and ravens just like each other's company. They have been observed playing a kind of game of tag. A group of ravens would dive at wolves and then leap, just in time, out of their grasp. (The Wolf Almanac) L. David Mech writes that, "Both species are extremely social, so they must possess the psychological mechanisms necessary for forming social attachments. In some way, individuals of each species have included members of the other in their social group and have formed bonds with them." These observations underscore the social adaptation of the wolf. His social skills may be applied to his relationship with man, and explain in part his selection toward becoming Canis lupus familiaris.

In Promise of the Wolves, Hearst creates Tlitoo, a raven, as the young heroine wolf's best friend. In the end, he conveys the message that will save Kaala's pack and avoid a deadly war. In the novel, the Greatwolves reveal a paradox: ". . .humans and wolves must be together but can't be together. . .humans need us with them to keep them close to nature so they don't destroy everything. But they fear us too much to keep us near and then we fight with them. That's the paradox." Humans would have to suspend some disbelief in order to buy into the first premise of the paradox that humans and wolves must be together. But, irregular as it sounds, that assertion might explain our almost mystical attraction to things wolfian, and our need to have canines as companions, an almost universal phenomenon. People who visit and support wolf sanctuaries are passionate about their connection. They seem thrilled to be in the presence of so many wolves, and find the howling in unison to be a gift that they are almost unworthy to receive. Everyone speaks of meeting Raven as a privilege.

To return to the ravens, the special friends of wolves, let us simply observe that, while humans may be said, occasionally, to run with wolves, ravens do fly with them. And, according to Mech, ". . .each creature is rewarded in some way by the presence of the other and that each is fully aware of the other's capabilities." Would that the same thing that is said about wolves and ravens could be said of wolves and humans. The wolf Raven was well on his way to making that mutual respect possible for countless people.

Raven's human family, notably the Cougars and close friends and fam-

photo: Phil Sonier

photo: Jan Ravenwolf

ily, referred to the stately wolf as "Fluffy." The oxymoron was just plain fun. This was a beast, tall and proud, black and silver, with a flash of white on his chest, whose only appropriate natural enemy was a grizzly bear, a wolf who could slay any of the amazed humans who approached him, with little effort. He answered surreptitiously, sometimes, to "Fluffy."

One afternoon in the woods of Candy Kitchen, Raven went by the name Best Man, when he stood by Leyton's side as his man and Georgia exchanged vows. Raven was also called black-phase timber wolf. This type of dark fur wolf starts out solid black as a youngster, and then as they age, white fur begins taking over. This process of going grey is known in the wolf world as, "Silvering Out." Perhaps we should all consider adopting this more distinguished title to describe ourselves as we undergo our aging transformation.

photo: Solveig Lange

photo: Jan Ravenwolf

Leyton Feeds Powder
photo: Georgia Cougar

# CHAPTER 10

## MEANWHILE, BACK AT THE RANCH

The founders of Wild Spirit Wolf Sanctuary, Jacque Evans and Barbara Berge retired. Leyton became the Director of the former wolf ranch. He had more wolf psyches and bodies to look after than just Raven's. The organization is non-profit and serves with a board of directors, a few paid assistants and several dutiful volunteers. They fend off dozens of calls asking for them to assume the care of a wolf or wolf-dog. They accept the animals when they know that there is room for them and money in the budget for their care. Some of their rescues are more interesting than others. Some are perilous.

Raven was getting to be an old man, as Leyton put it. Wolves in the wild rarely lived until age seven. Raven was still serving as ambassador until he was thirteen. He had the same maladies as older people. He would get down in the hips, not be able to jump up into the van (not that grandpa has done that recently), find it difficult to stand and sway with the movement of the van from his position in the back. Standing and visiting with mobs of people was getting to be just too much to ask of him. Leyton began thinking the inevitable; he would soon need a new ambassador. Raven could not be replaced, but he should begin at least part time retirement. Which among the wolves could even begin to do what he did, so gracefully, so dependably? Most of the other wolves did not take so kindly to humans-usually with good reason. If they were rescue animals, then they bore scars of every kind.

Leyton began calling contacts in the wolf world to see if any wolf pups were available. Taking in wolves that were not for rescue, did not fit the sanctuary's mission statement. However, Leyton was searching for a special wolf, a new ambassador wolf. He located a breeder in Oregon whose pair of arctic wolves had given birth to six pups. Four had been spoken for, but Leyton could have two. He quickly accepted the offer. He thought that if the two little wolves were bottle fed and raised by humans, they might have the psychosocial qualities to be among people as ambassadors. Leyton was to take his family on this long road trip. Just before they took off, the man in Oregon called to say that they could have all six of the pups. He said that he was tired of the potential owners' hesitation and hassle, so Leyton could take them. He jumped at the opportunity, thinking that surely out of a litter of six; one might grow to have that special disposition that Raven had. The car wasn't really prepared for that many, but they thought that they could manage. "Managing" is what they do best.

So, Leyton and Georgia and Lakota drove back home from Oregon to New Mexico with the equivalent of sextuplets. The wolflets had to be bottle fed, cleaned up after, entertained and, I guess, burped. Their eyes were barely open, but they were capable of constant noise making. They were all little grey balls of fluff. They were the offspring of Sierra and Yukon and were named, either on the trip home or shortly thereafter: Storm, Sugar, Sabine, Thunder (also called Bob), Alice and Powder. They began to be raised very carefully. For a time, two or three were brought to sleep with the Cougar family. Angel slept with them all for a time in one of the offices. The Cougars and Angel alternated sleeping care. The pups were protected from parvo by a disinfectant; there was a lot of hand washing, not to mention hand wringing. Georgia says that the wolf pups do not feel, much less act, like dog pups. They don't feel soft and fat, rolly polly. They are muscular and wiry; their teeth are sharp and their fish hook (Georgia's word) claws begin operation soon. All of the caretakers had bite and claw marks a plenty. Yes, they were cute. I had a picture taken holding some and I look goofy, not to mention what I sound like making baby talk to wolves, tiny or not.

Liz Parr with Pups
Photos: Georgia Cougar

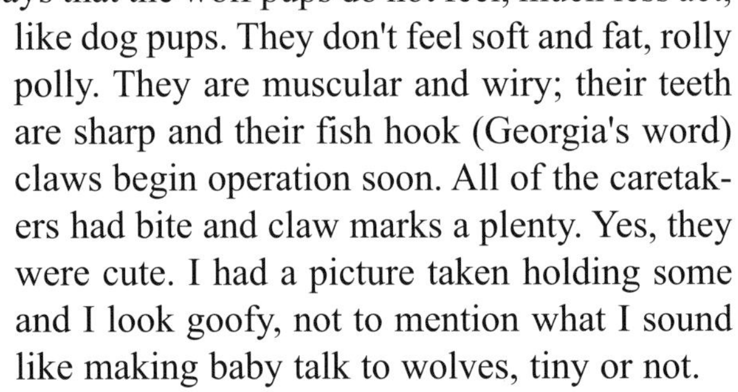

They grew, as expected, to be gangly "teenagers." They are now not yet mature, but Leyton has not found a young Raven. He says that one really cannot judge them at this stage of growth. When they are mature, they become wolves. By that, he means they may go through a real psychological change, with the inbred impulses surfacing, with an innate dislike for the different, including humans, and with their constant, alternating attempts at being the pack leader. They may not be gentle at all. Leyton is Dad to them. He is the bearer of authority, but even he does not know what to expect when they reach maturity. Just within this past month, Leyton and Georgia were at the sanctuary office when Leyton thought that he heard something amiss, and had a feeling that something was in the air. He went to the enclosures to check. As he drew nearer, he heard the snarling, growling and squeaking that signaled trouble. A couple of two-year-old pack mates had cornered three-year-old Sassy and ripped into her. Some flesh was hanging. Georgia heard Leyton roaring and thought of the time that Leyton had entered a wolf enclosure to find Nikki on the attack. Nearly exhausted, Leyton managed to get Sassy away from the two. He quickly determined that she needed to be rushed to the vet where she was treated for battle injuries which included a punctured lung.

A year after the first installment of pups, the breeder in Oregon called again. He said that he was giving up wolf breeding and asked Leyton if he wanted to come and get Sierra and Yukon, the parents of the six pups. Leyton was delighted with the news that the man was going out of the wolf business and said that he would. This time, staff member Angel went with Leyton in the van. They were better prepared, and determined to drive straight through there and back, alternating drivers. Now, this was a real rescue and more in line with what the sanctuary does. The only drawback was that they had to catch the sire, Yukon who, other than exercising his job of breeding, ran wild. Sierra was already expecting.

Finally Yukon and Sierra made the trip home to Candy Kitchen, New Mexico. They could retire in ease at this lovely sanctuary, only a few yards apart from their last litter of wolf pups. Surely one wolf, out of two litters would grow to be a workable ambassador, a nascent Raven. There was to be some misfortune involved in this family pack, however. Sierra gave birth to five pups at Wild Spirit; the staff decided to let her keep and nurse three of her pups, while they socialized the two others. They were all named after either mountains or the

Sierra & Yukon photo: Allison Bailey

Sierra and Her Baby    Photos: Allison Bailey

snowy caps a top them: Teton, Trinity, Shasta, Frost and Flurry. Sierra was able to begin to raise her pups for the first time in her life. Among breeders, who sell this wild breed to all comers, the regimen is to remove the pups from their mom by the tenth day. However, not long after Sierra arrived, she died from a combination of heart failure and complications from a Cryptococcus fungus found in the soil of her enclosure. In the fall of 2008, Trinity succumbed to cardiomyopathy. A vet decided that Sierra and her offspring might suffer from a genetic defect. At this writing, however, in the fall of 2009, the brothers and sisters are thriving, and it is Flurry who seems to have tendencies toward ambassadorship.

Considering Flurry's behavior, Leyton has seen to it that his enclosure adjoins Raven's so that they can talk through the fence. Leyton has admonished Flurry to listen to the old man Raven. He will tell Flurry that it is much more fun to go with Dad on adventures than to stay and be bored within enclosures.

Leyton had begun his adult journey with the words of a teacher, "Wolf medicine is strong medicine." At the time, he understood nothing of the prophetic words. He stepped into Native American lore on his way to a commitment to one of the sacred spirits, the Wolf. Barry Holstun Lopez in his lovely book, Of Wolves and Men (Scribner's 1978) writes of the relationship between wolves and warriors in the Native tradition:

"The wolf fulfilled two roles for the Indian: he was a powerful and mysterious animal, and was so perceived by most tribes; and he was a medicine animal, identified with a particular individual, tribe, or clan. In the first role he was simply an object of interest . . At a tribal level, the attraction to the wolf was strong because the wolf lived in a way that made the tribe strong: he provided food that even the sick and old, could eat; he saw to the education of his children; he defended his territory against other wolves. At a personal level, those for whom the wolf was a medicine animal or personal totem understood the qualities that made the wolf stand out as an individual; for example, his stamina and ability to track well and go without food for long periods. That each perception contributed to and reinforced the other-as the individual grows stronger, the tribe grows stronger, and vice versa-is what made the wolf such a significant animal in the eyes of hunting peoples." (Lopez)

Leyton tells of a humorous encounter between Raven and one of the wolf pups. Raven had just eaten in preparation for a long trip in the van and another presentation. The pup came sniffing around Raven. Apparently there was some exchange that indicated that the big wolf should feed the little one by regurgitating his recent meal. Raven tried his best to keep his food down; he swallowed in an effort to retain his lunch, once, twice. . .and finally had to follow the course of his nature. Wolves seem to follow very easily the suggestion that it takes a village, in this case, a pack or tribe, to raise a child. The pup waddled away, full and happy.

photo: Georgia Cougar

Photo: Jan Ravenwolf

# CHAPTER 11 

## DO NOT GO GENTLE INTO THAT GOODNIGHT

As often as these lines, from the Dylan Thomas poem, have been quoted, I can think of none better to describe Raven's passage to the spirit world. His light was dying. Those who saw him often commented on the diminished light in his eyes. His eyes were memorable, amber, like most timber wolves, blazing sometimes, soft and welcoming at others. Jan recalls a time when, at a presentation in an auditorium, Leyton and Raven on the stage, she locked eyes with the wolf and walked, as if mesmerized, to the stage where he stood. But now, near the end of his graceful, generous life, the light was fading. Helen remarked on that on his next to last visit to Grandma's house. He still performed the ritual that went with the visit, the wash cloth cleaning, but he turned down the pieces of pork that Grandma had cooked for him. And when he left, Leyton had to pull the van next to the curb so that he did not have to jump up to get in. He left me with a gift, however. I was seated at the kitchen table, so eye level with him. Without any coaxing, he laid his head on my chest for just a moment. I thought later that that was a goodbye gesture. Yet, I was very far down the list of his familiars.

A few weeks before his death, Raven was let out of his enclosure to walk the sanctuary grounds freely. One day he couldn't find his way back to the gate that would let him

photo: Jan Ravenwolf

in to his enclosure. This the sure-footed wolf who had once guided Leyton through the labyrinth of a contemporary school building, the prince of the wolves who had walked past the enclosures with his massive head held high, checking on the subjects, requiring their attention.

On the Saturday before Easter Sunday 2009, Leyton and Jan intended to take Raven out for a walk, but he didn't seem interested, didn't want to put the leash on; he just sat in one spot for a long time. It was snowing. They got blankets to cover him. It began to snow more and more and someone brought Leyton a blanket. Everyone at the sanctuary began to come around; Lakota was there to comfort her Dad.

Photo: Georgia Cougar

They all talked with Raven for awhile. Once when Raven stood, in an effort to breathe more easily, Leyton and Georgia managed to place a blanket under him, so that with others assisting they made a sling to carry him into the sanctuary office where it was warm and dry. Raven did not like that one little bit. When Leyton had tried to maneuver him inside the first time with the leash, he had turned and bitten down hard on Leyton's arm. He was wearing a thick jacket, but even with that, he thought that Raven had broken the skin. It was the same arm that Raven had chomped on when they first met. The female staff had fixed a bed inside, cushy and soft. It seemed that he just wanted to curl up and be left alone. He began to drink a lot of water and then vomit. About midnight, Raven had stopped vomiting and started seizing. He seized seven times. Angie, another close friend of the sanctuary, and Jan were present with Leyton, who says that Raven seemed to struggle to die. Every time he seized, he cried. Angie and Jan took a break leaving Leyton alone with his partner. Just at sunrise Easter morning, April 12, 2009, Raven, the ambassador wolf to humans, died, his head in Leyton's hands.

photo: Georgia Cougar

#  EPILOGUE 

You can't make these things up. Raven really did die at sunrise on Easter morning. The wolves howled in unison. A blanket of snow covered the grounds of Wild Spirit Wolf Sanctuary. A devastated Leyton took Raven's body to have the pelt preserved. That was their last journey together. Now, what is next?

First always are the ordinary activities. The staff must prepare what is called a wolf loaf, a careful combination of healthful ingredients for each wolf on feeding days. The quarterly newsletter must be written, edited, printed, collated and mailed to thousands of supporters who will read of Raven's death. The envelopes with donations will come in, the people at the sanctuary hope, for otherwise the wolves would not be fed. Fencing for enclosures will be mended, expanded. There is talk of a memorial pond for Raven. Before long, another of the senior citizens at the sanctuary will pass away and the wolves will howl again in dirge for the departed. The board members will brainstorm over money sources so that the non-profit work can continue. The staff will continue with their less than ideal tasks, things like cleaning out the enclosures, watering and feeding the wolves, while they often do not show any appreciation. Mostly they growl with that barely heard rumble that seems to come from someplace under them-the wolves, not the staff. Well, sometimes the staff. Then there are the constant trips to talk with people about what they do, to try to sell the souvenirs and T shirts that bring in some revenue. All this goes on while they are in the

Angel with Forest

photos: Jan Ravenwolf

Flurry with Friends

scary space of having no ambassador wolf.

Some of the teenagers are showing promise, Angel's Forest, that she practically raised, is good with her. Leyton had a few good appearances with Storm. Recently, Flurry appears to be the most suited to the patience required, and he likes humans well enough. He tolerates us mostly. Leyton says that Flurry hasn't a clue to what he is about, whereas Raven always knew. Raven came to Leyton at two years of age. Most wolves don't reach any kind of maturity until about four. Raven was grown up at two, according to Leyton. Of course, Raven did go through that four month period of sulking in his tent [enclosure] like Achilles. There will be that period, which mere mortals afford the great, of thinking and remembering that they did no wrong. These little teenage wolves are all beautiful; the arctics like Storm and Flurry have snowy, deep white coats; the timber crosses like Forest, are more tan and grey. The public's expectation at a wolf outing is just that they get to see one. The wolf need not be charismatic, as long as Leyton remains so. As he says, he is the voice, the wolf has the looks.

Leyton expects eight new rescue additions soon, three wolves and five wolf-dogs. The wolf-dogs are a family pack. They have been adopted out and then returned so often that Leyton intends to keep them together. They seem to get along very well that way. And so the sanctuary begins a new era. This era is built firmly on the partnership of Leyton and Raven.

Together they sent out the message that all life is valuable, that no living thing should be subordinated to another carelessly, that no life form is here to be neglected or exploited, that living beings can co-exist if only one does not dominate selfishly. And each should be allowed to be what it is. The Spirit of Wild Spirit remains with the Sanctuary.

*Hey, you, under the tree there, can you hear the not so distant howl?*

I don't even know how or where to begin. I just found out today about what happened to Raven. I cannot even find the words to type, though I'm sure you know how much Raven meant to me and to my father as well. I am so thankful to you for taking me into his enclosure to play around for a bit while I was there a few weeks back. It brought back many memories and reminded me of just how special he was.

I can't even imagine what you have been through over this time. Raven touched the heart and souls of any who were lucky enough to have him cross into their path. But for you it was different, it was deeper than that. The two of you were one in the same. Kindred spirits. Brothers. Best Buddies. No one could speak of Raven, or tell his story, without including the special bond the two of you shared. Raven may be gone now, but that bond will never weaken. His legacy will live on forever.

I am sorry,
Fred Davis

I am saddened and dismayed that I was not able to see Raven the last two years. It is difficult for me to type with tears streaming down my cheeks. Leyton, I am so glad that you were with Raven at the end as I'm sure it was comforting for both of you. I am sure there will be other great wolves--but there will never be another Raven. His lovely and memorable howl will be sorely missed. Leyton, you are the greatest.

Sincerely,
Bernice Dalby

Dear Leyton,

We are very sorry to hear that you lost Raven. It is good that you could have at least been there with him, comforting him and telling him farewell, when he was leaving. It must have been a very sad moment in your life to see him go, but it really means a lot for both of you that you were there. Raven will never be really gone. He will live on in your heart and the hearts of many others that he touched. He was truly an ambassador for his species building connections of his kind with humans. I will always remember how I howled with him to make all the other wolves howl on tours. We feel with you and will miss him too.

All the best for you and every creature at Wild Spirit Wolf Sanctuary and we look forward to seeing all our friends again when we come back to the Sanctuary.
Solveig & Kris

With a heavy heart I want to send my condolences to you and to everyone at the Sanctuary. Losing Raven must have been awful for you. Raven was such a sweet, loving and beautiful wolf. I met him 3 yrs. ago during a visit to New Mexico . He touched my soul that day. When he started to howl and got all the other wolves howling, it made me cry. I was so HAPPY to see wild wolves to hear a wild wolf, it made my heart skip a beat.

I sit here, remembering that great autumn day, meeting such an amazing animal. Wondering WHY others can't see what I see? That the wolf is a beautiful, soulful, amazing animal - that NEEDS to be part of this planet. We need to protect them.
What you do at the Sanctuary is truly awesome!!

Having Raven all those years must have been a joy. He truly was one remarkable wolf! He is now part of the big wolf pack in the sky. Howling & roaming freely. I feel blessed that I did get to meet him!! I will never forget his howl, so sad, yet so beautiful, it has stuck with me ever since.
God bless!
Jackie Palmisano

*Photo: Allison Bailey*

*The Author at Home with Raven* photo: Georgia Cougar